Copyright© 2021 by Cytrine Buczko

All rights reserved. No part of this book may be reproduced or used in any manner without written permission of the copyright owner except the use of reprints in the context of reviews.

VANISHED
BUT STILL HERE

POETRY
STORIES

Cytrine Buczko

To all who are lost
and all who are found

Contents

Time poetry	9
Fade & Appear poetry story	12-84
Four people poetry	88-89
Too far poetry	90
Unknown poetry	91
Life poetry	92-93
The farmer poetry story	94-103
Missing beauty poetry	107-110
Sphere poetry	111
Maybe one day poetry	112

TIME

Tick
i am here
tock
i am gone
tick
i appear
tock
i move on

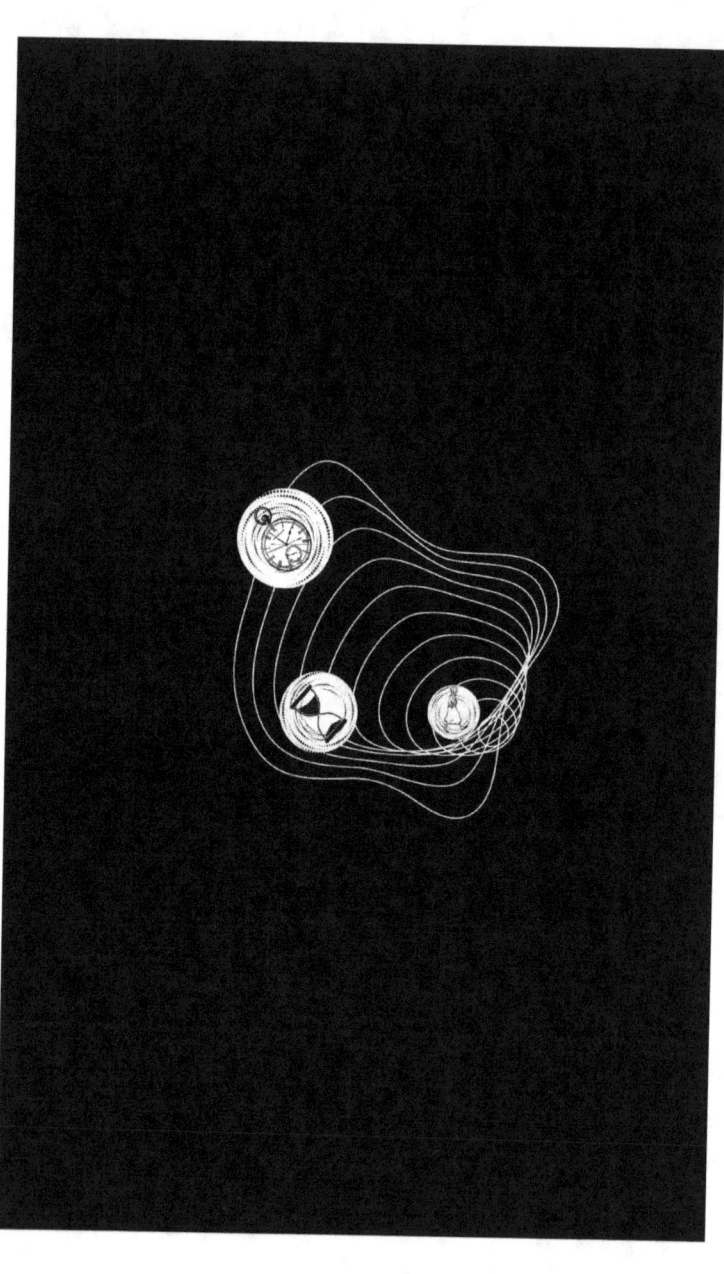

fade
&
appear

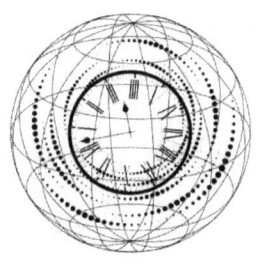

poetry story

WHAT HAPPENED?

Stones eyes a crystal
his smile a moon of love
i grin
he steps closer
his strong hands
on my cheeks
his breath so near
our lips meet
i close my eyes
yes
he is the one
forever
i will get some drinks
make sure the fire stays on
he raises
i watch him
his body
his soul
so good
so perfect

don't leave me
he says
as if he really fears
never
i answer
settling into the bench
in our new backyard
i love you to the moon
Elly Whitlock
his words a song
he turns
i watch him walk
towards the house
our house
a long awaited goal
a dream
a wish come true
i take a deep breath
i close my eyes

silence
this house
surrounded by green
close to the coast
winter almost upon us
i love it
it belongs to us
idyllic
so quiet...

I wake
silence
it surrounds me
so quiet
my eyes squint
moonlight
bright
a full moon
i feel different
a nice nap for a change?
as if i had slept
a hundred years
my body off balance
i raise
my eyes scan
where am i?
my boots sink into snow
deliciously soft
first snow?
how could i have missed it?

it was autumn
just a while ago
i fell asleep
on this bench
now i see winter
so soon
so abrupt
i am confused
i shiver
in my knee long dress
damn unpredictable weather
i wonder
where is Stone?
he had left the bench
to get me a drink
why wasn't he back?

i feel lost
him
me
our hearts
always together
i turn
i ask
Stone?
no answer
silence, still
what happened?
did something happen?
or is this just
imagination
a dreamlike state?

i hover
i stare at the bench
i realize
a different bench
not the one Stone left me at
not our yard
not our house
this is different
everything
i don't know this path
trees look unfamiliar
my eyes catch a glimpse
a house
not far from this walkway
i shiver again
my mind spinning
how did i get here?
i raise
my steps heavy somehow
slow
and then faster
it's snowing now.

A house
a door
my knuckles
hit the wood
i wait
the door opens
i feel out of place
a man stands
his clouded eyes waiting
for me to speak
where am i?
my words are brittle
Stout street
his voice old
is the coast near?
he laughs
this is backcountry dear

my eyes scan around
no ocean?
he shrugs
9 hours from here
i am not home
unfamiliar
everything
where am i?
i turn back to him
a desperate begging
i need help
he nods
come in dear

Colors flash
my eyes hurt
the lights are too bright
surrounded by night
i watch police
they just arrived
a hopeful thought
Stone could be with them
he is police.

A man approaches
an officer
not Stone
Eleanore Whitlock?
i nod
his face crayon
button eyes unsure
you need to come with me
he turns
i follow
reluctant but trusting.

A car
warmth
the only comfort
the journey is long
no one talks
i try to ask
where i am
how i got here
no answer
just a
*well get to that
when we arrive*

i am tired
my body slumps
i close my eyes
no don't fall asleep
not again!
i force my lids open
i can't...
i can't stay awake.

The policeman
he wakes me
i look outside
my eyes now burning
light
it is morning
how did i sleep so long?
i recognize the town
my town

the car drives slowly
as if to use caution
a precious cargo
me
inside
confused
but happy to be home

almost there
he says.
I wonder where
maybe my house?

i don't ask questions
he told me
he can't answer
i don't understand
but i accept

we arrive
Klover Valley
small
cozy
ten minutes
from the coast
my home

the car stops
police station
that's where we are
i follow the cop
we enter the building
from the back
like a pop star
almost secretive

a deep breath
we are inside
it is darker
my eyes drawn
to the darkness
i wonder why;
light
never a problem
never a bother
for my eyes before.

We walk down a hall
the smell
concrete and coffee
my hands shake
i want to shower
i want to go home
i want to see Stone

we enter a room
interrogation?
my eyes sweep the space
bleak
sit
i comply
why?
i don't know
i am tired
so tired

a few minutes pass
i am alone
my mind is spinning
why am i here?
slowly i feel
i need to call Stone
he is a policeman
he is my boyfriend
almost 4 years

we have a house
we have a life
friends
his family mine
my father his
i want to call him
or call my father...

the door opens
people walk in
i don't know them.

Questions start
about where i have been
confusion sets in
i don't understand
now they ask
what day it is
what year it is
i answer
honest -
silence
i look at them
they look at me
one shakes his head
then they talk
explain to me
i listen.

VANISHED

My heart pounds
i falter
i fall to the ground
and don't get up
a hotel room
for now
they said
for i have no home
i am here
my body
on a dark blue carpet
alone
scared
painful heartache

they brought me here
after they talked
after their unlikely
story
so strange
it couldn't be real
but somehow it was

their story to me:

i vanished from home
that night Stone left
to get me a drink
i disappeared
thin air they said

everyone searched
everyone helped
everyone cried
everyone blamed
one year
then two
no sign of me
a funeral
and with his heart
broken apart
my father's life
ended
a year ago
his grave
nestled
next to my mothers
next to mine

and Stone
oh Stone
a tragic fate
scolded by haters
called a murderer
his time in jail
as a killer of mine
a win for the ones
who believed he killed me
a loss for all family
for friends
for my father
who believed
in Stones innocence

now in prison
already past a year
with no body found
no guilt
but yet branded guilty

my return
will set him free
but our love
the one we shared
with deep desire
and thoughts of marriage
that love he now hates
i am sure of it
breaking my heart
breaking me

RETURN

i work
a small cafe in town
eyes on me
always
everyone knows
rumors floating
my memory blank
my time away a mystery
everyone asks
including me
am i a ghost?
do i know things?
was i abducted?
lived amongst
the unexplainable?
amnesia maybe?

when i woke
at the bench

my dress was the same
my hair in curls
like the day
i disappeared
there was no sign
i had ever been gone
but miles away
from where i vanished
and wearing my boots
my earrings
my necklace
how can i not be
a stranger case?
for people
police
for friends
and family
so eery
so incredibly frightening
everyone baffled
including me
everyone almost scared
of me

people search drama
their heads together
asking
am i a liar?
setting up him,
Stone Kane,
to go to prison
so dangerous
him
as a policeman
between the guilty
a minefield
for him

why would i ever
do such a thing?

their reason
a made up one
they say
he wanted
to leave me
i know he didn't
i know him too well
and even if
to see him in danger
in a prison
as a cop
i would never
do such a thing.

I shake a lot
my body
my hands
anxiety maybe
i don't know
prying eyes make it worse

i quit my job
i start as a maid
a small hotel
outside of town
no eyes on me
i can hide.

One month
then two
i feel out of place
doctor's visits
a lot,
too much,
my blood is drawn
for safety they say
examinations
head and body
i let them do it
to not create friction

although i hate it
all these questions
as if they search and hope
for extraordinary
for different than human
for strange dna
of course they don't say it
cautious they are

they ask about
a scar on my wrist
i know it is new
but i tell them it's old
i keep it a secret
i am afraid
they might keep asking
maybe take me
to do more tests
not leave me alone
or think
i am an alien kind
if i tell them
it's new.

My thoughts with Stone
i crave his presence
but no sign of him
no effort to connect
me
him
both in silence

i know he is free
i keep telling myself
that's all i want
him free
him happy
him not with me.

Routine
a forced motion
a small apartment
near the hotel
my auburn hair
chopped to my chin
button eyes
never in makeup
my figure
not healthy
i can not eat
my mind is spinning
every day
the same question
inside my head

where have i been?
two years and two months
that time has vanished
just like i did
that time is timeless
no part in me
no part of me

did i grow older?
i am not sure

a psychologist
assigned to me
sessions
talks
notes taken
after weeks
i still am
a mystery
i still am
out of her league
she can not help
she can not comfort
i quit her
i am alone now.

Reporters surround me
like crows
pecking at my heart
calling me alien
calling me fraud
calling me a demon

people
siding with Stone now
pretending to know
that i did this
on purpose to him

no friends
no mercy
no simple life

my mind racing
thinking of moving
living alone
off the grid
but too afraid
i can not be lonely
what if i disappear again?
a city maybe?
with millions of souls?
i wouldn't be happy
my heart in nature
not concrete jungle
where should i go?
where should i turn?

A GLIMPSE

Threatening letters
my door smeared with hate
uneasy i go about my life
i see people hissing
avoiding me
looking at me
like a witch
on trial
i can not explain
i can not prove
to be anything other
than what they say.

Weeks pass
i walk in darkness
my thoughts still hover
who am i now?
who was i then?
my life so different
so lonely
so desperate

two years ago
friends showered me
my father always there
a listening ear
a saving grace
my cousins now shy away
they refuse me
maybe they are scared of me

and Stone
once upon
a time ago
a soulmate
always there
protecting me
strong
gentle
the most gorgeous of men
like a layer
around me
loving me dearly
now silence
no word
no sign

US- STONE AND ELLY

us
first as friends
in school
sweet playful days
before he moved
us kids
saying goodbye
to each other
time passed
and like fate
as adults
we collided
in the unlikely place
of a minor accident

where he was the cop
i was the injured
he saw me
i saw him
without doubt
a lightning struck
our hearts together
ever since
and- so we thought-
to infinity.

We bought a house
a garden
a home
shared friends
shared family
shared friendship
so good together
years of bliss

I cry
my body on the couch
my soul not here
i hold a bottle
fermented grapes inside
but can not drink it
afraid
my total numbness
would lead to more,
more bottles
more booze
to forget
who i am now
a vicious cycle
too easy to achieve
so dangerous to start.

A sound outside
i hover
another sound
right by my door
my heart beats fast
fright enters
i know it could be
a hater
a conspiracy believer
alien hunter
it could be police
government
the press
or just a cat
i shouldn't worry
but i do

i lower the light
stare at the door
i take my phone
technology
a helpful tool
a camera outside
to spy on people
who surround the door
alerts me
someone is here

racing heart
i wait
commotion
a shuffle
a thump
a scratch

then silence
nothing more...
what now?
i wait
i can not move
but then i do
i take my phone
i quickly push
the playback button
to see
what happened
who is out there
i watch
a man
short but strong
his face covered
a mask of some sort
i shake
terrifying to see

i watch
the man
approaching my door
my breath is gone
my heart is racing
the man
in mask
he pulls out a tool
something to enter
something to break my door
my eyes glued
to the video
the man attempts
a breaking of sorts

a sudden change
his attempt interrupted
a second man
out of nowhere
attacks
they wrestle
a chokehold

the second man
taller
although it's dark
i recognize
his body
his movements
his way of fight
precise
professional
strong

Stone.

My heart races
Stone in advantage
overtaking the man
whose intent is harm
to break in
hurt me
no doubt about it
Stone now
dragging him away
out of sight
my camera
a witness to a rescue
my mind spinning
everything quiet now
no intruder
no Stone
no sound.

Tears
wine clenched in my hand
Stone
a dedicated policeman
looking for the good
taking care
of everyone
always
today a guard for me
my heart drips
i falter
i can not deny my yearning
him protecting
me
does that mean
he cares?

cares about me?
the one
who had made him a killer
the one
who had sent him to prison
the one
who had left him
for two years
with no word?

he is still here
fighting off strangers
at my front door

  why?
  why was he here?

STONE

Days pass
i ignore the resilient haters
my work a damp
dark place
no end in sight
no surrender of my anxiety.

I hold a sunflower
hard to find
in winter days
but my father's favorite

my boots in the snow
my eyes set on him
my father in name
engraved in stone
under earth
his soul long gone
now next to my mother

me
desperate to tell him
that i am okay
that i am back
alive
not dead
here again
but now without him
i break in silence
a wave of tears
a father's love
so powerful
in my heart
thoughts of him
a proud man
a wonderful man
without him
i am lost
he is an angel
now soaring
in all the kindness
all his beauty

i take a deep breath
the air peaceful
around the graveyard
a feeling of comfort
maybe because
my father is here
my mother is here
i am with them
in thought
in soul

the stillness
now suddenly broken
by sounds
a click
then two
reporters swarm me
i don't understand
i can't catch my breath
why are they here?
bodies closing in on me
i drop the flower
i lower my head
my sunglasses protect
just a little

questions raining
again and again
of encounters
of the unknown
why again?
i thought it was over
why are they pushing
i can't comprehend

they seek out
the most vulnerable moments
to hack
to blame
to find a reason
i can not move
i just stand
in front of the graves
i don't know where to go

Elly quick
a voice so strong
between flying words
between clicking
and pushing
and rumors
of aliens
of dna
of lies
deception
between pleas
for my side
of the story

i feel a hand
strong and determined
taking my arm
pulling me through
toward him
ah yes
my protector
my guard is here
Stone
his tall body
a shield
he covers me
threads us around
the flock of ravens
still hacking
while he wraps his arm
over my shoulder

he leads me to his car
reporters follow
no mercy
no dignity
no sense of right
get in
we'll get your car later
Stone's voice
so calm
but still fierce
so familiar
my heart hurts

he drives fast
but also cautious
silence slicing
no word he speaks
no word i say
wherever he takes me
i trust he cares
about who i am now
about me

he navigates
shakes off reporters
until it is quiet
a side street in town
we pull in
to hide for a while
dark
the afternoon light
not strong enough
to fill the walls
surrounding us
he turns off the car
we don't speak
my heart a mess
my eyes fixed
onto the windshield
in front of me

i can not look
in his direction
i am afraid
afraid he changed
afraid of what he sees in me
afraid of certainty
that i have lost him
lost his love

Elly
his deep voice
the way he speaks my name
a shower of total yearning
so familiar
so loving
so hot

i shake my head
and lower it
my fingers
wrap around my wrist
the new scar
somehow it brings comfort
when i touch it

please look at me
i hear him beg
i rub my wrist
i close my eyes
sunglasses still on
like always
ever since i appeared again

i can not speak
or maybe i can
i am not sure
not sure of anything
my insides like clutter
a tug and a push

my heart is racing
i take off my shades
and with a deep breath
i raise my head
to look at him
and just like that
i fall into pieces
a dark hole
total pain

his face
a scar above his eyebrow
made the rumors true
fights in prison
they taunted him
for what he had supposedly done
and him
a policeman
dangerous to be
surrounded by murderers

vicious fights
probably unfair
one against five
and the scar
a reminder
i put him there

my honey eyes
fall into
his crystal blue
that gaze
a magic in it's own
damn handsome
with his raven hair
strong jaw
his lips

his hand quickly brushing
over his mouth
a deep breath
and then his voice again
God Elly
i really thought
you were dead

i feel his relief
i see it in his eyes
my agony
my guilt
a constant visitor
in my stomach
i quickly tear away
and look down again
wet eyes
while i say
i am so sorry
i am so sorry
for everything

no
don't say that
it was not your fault
whatever happened
it was not your fault

my breath stops
does he trust
trust i tell the truth?
i am not sure

the car is silent
a drop of tear
now running
over my cheek
as he leans back
into his seat
i need to talk
say something to him

i fell asleep
then i woke up
i saw you
just a few weeks ago
that's all i remember
that's all i can tell you
i wish i could
say more to you

*i owe you something
an explanation
i am the one
who left you
sent you to prison
two years
where did they go?
i am not sure
and for all this
i am
and always will be
sorry*

my ramble
mixed with tears

*Elly i believe you
i know you don't know
you don't know
where you have been
i see it
in your face
your sadness...
i researched
i wanted to know
about your case
and i found others
there are other
cases like yours
you are not alone
i found them
i've seen them
talked to them
they are here for you
i am here for you*

i look at him
he looks at me
a sadness
his crystal eyes
a flicker
he scans my face
he shakes his head
to rid his coming tears

Stone

my whisper broken

his eyes
now drenched
he whispers

you are still the one
i still yearn
to have you by my side
i know
everything is different
but i am not
i swear i'm not

my heart beats fast
a feeling
of hope
first time since i woke
stillness
now broken by tears
his eyes
a reflection
of two years of agony
a life without
each other

he cries
i cry
i climb on his lap
we embrace
a deep
beautiful hour
a dive into us
again
for him
after two years
for me after a few weeks
but still
like nothing ever happened
i drown in his smell
safe in his arms
his lips a sea
of everlasting emotions
i touch all he is
i feel his pain
we fit
like we always have.

NEW

Weeks fly by
our hearts
together
our minds one
he saves me
i save him
a beautiful thing
we want each other
to strengthen
who we are now

without him
i would have fallen
deep and alone

with him
i am fighting
i am flying.

We move away
this town
not ours anymore
a new home
beautiful trees
small valley
mountains
rivers
new people
they don't know
my strange story

we connect
with others like me
their tales
as strange as mine

some vanished
just a day or two
some for a year
no memory
children gone
then found again
impossible places

no recollections
no concept
of how they could master
a distance so far
a time so short
how they survived
one year somewhere
alone
or maybe with someone
how they lost their minds
their memory
no clue
of where they have been
and others,
members
of a loved one lost
in thin air
no sign
no trace
just gone
from one second
to the next.

I build a life
again
Stone by my side
our children the future
our love an endless bond

my scar
a reminder
of what had happened
but my life
a reminder of
where i am heading

my friendships with others
our mystery combined
who we are
where we went
we can not say
but bonds are formed
and a life is carved again

and maybe
one day
i will remember
where i have gone
vanished to
and appeared again
i will remember
why i have this scar
and maybe one day
i will tell the story
of the two years lost
two years
somewhere
nowhere
everywhere...

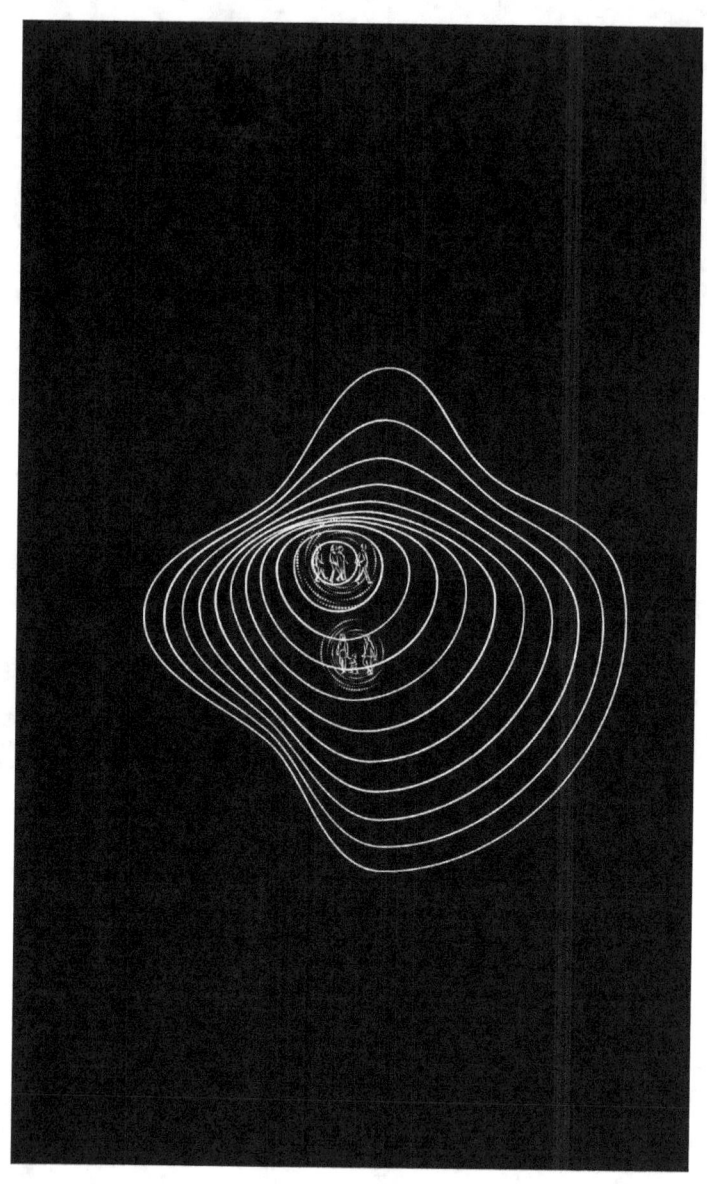

POETRY

FOUR PEOPLE

Four people
a path
four people
a brilliant day
trees
sunshine
and clouds
familiar routes
great hike
four people
no stress
fun words
friendship
four people
a hill

one person
over the hill first
and suddenly
only three left
three people
searching
desperation
one is gone
no one knows
disappeared
no sign
never again
thin air

three people left

TOO FAR

A toddler
gone in a second
found in ten minutes
lightyears away
all on his own
all alone
too far
not even close
how?
why?
mystery remains
no one knows

UNKNOWN

The thing about the unknown
it rattles us all
it finds no mercy
it just keeps going
it is sly
and smart
it brings us to our knees
a small bite
poison
or paradise
the unknown clings to us
holds on
sweeping mad
and so intriguing
a great mystery
that drives us mad
but lets us live

LIFE

Life
a working magic
a mystery in itself
weaving in
weaving out
constant push
constant pull
directions you see
and then you lose
directions you find
and then you follow

life
a conscious struggle
and deep love
a question in all
rushing inside you
rushing around you
where are you?
where am i?
who are you?
who are we?
a purposeful stop
on a road ahead
life

the farmer

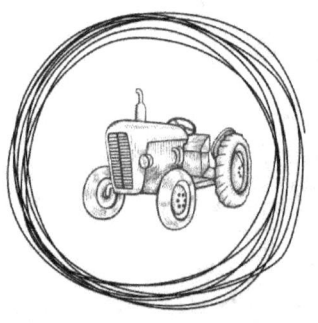

poetry story

TRACTOR

A farm
surrounded by land
a tractor
blazing sun
fields everywhere
high noon
working hard
a farmer
dedicated to his crops
the day
like always
started early
now break time
but the tractor is still going

the farmer's wife
she calls for him
to come on in
to cool off
lunch is ready
in this heat today
he waves
a sign he understands

the clock is ticking
the tractor still going
the ice is melting
in a glass of water
she has ready for him

the farmer's wife
she goes back out
to call him again
she watches
the tractor keeps going
her eyes in question
she can not see him
a strange feeling

down the porch
her steps are fast

she approaches
the tractor
calls out his name
now worried
the tractor
almost crashing
into the fence
she yells
she sees no man
she must get closer
he has to be there
on the tractor
now she runs
again
calls out his name

the tractor crashes
she stops at the site
a search for him
the seat is empty
the tractor is empty
she runs around it
he is not here
not on the ground
not anywhere
where is her husband?
where is the farmer?

SEARCH

Police arrives
a search begins
wide open fields
for miles stretched
surely easy
to find him close by
the heat ablaze
but still they look
he has to be here
no chance of disappearing
the fields too wide
too open
too far
certain
they will find him soon

bewildered
the search not fruitful
the farmer gone
no sign of him
he vanished
left with questions
his wife distraught
confused at best
i saw him
called out to him
to let him know
that lunch was ready
ten minutes later
he was gone
she tells police
she tells her son

so much pain
and so much searching
how can this be?
the heat that day
the fields stretch miles
the farmer old
not very sporty
and still
within ten minutes time
he disappeared
never returned

loved ones grief
for his lost soul
drowning
never ending questions

the search goes on
in hopes of something
the wife in pain
the son determined
to find his father

years pass
no father
no husband
no farmer
his disappearance
a mystery
never solved
like so many...

POETRY

SPHERE

Spooked like a deer
you disappear
into a sphere
taken by a puppeteer?
deep fear
falling tear
unclear
or
are you still here?

MISSING IN BEAUTY

It is the clean air
that we are drawn to
the open spaces
smell of the forest
the beauty of nature
a quiet place
that lets us breathe
a time to reflect
a time to relax
to come back home
refreshed
fulfilled
and with new meaning

rivers and lakes
forests
mountains
national parks
safe spaces
memory lanes
and still
places mystery finds itself

places some go to
and never come back
places they stay
maybe by choice
or vanish
strange unsolved cases
stories so odd
stories weaved
in uncertainty

a child
a man
a woman
vanished
tracks seen
but then gone
disappeared

clothing
found in settings
impossible to get to

memory loss
dementia
or something else

areas so beautiful
so calm
a wonderland
a lifeline
nevertheless
mystery still surrounds
and questions
about the missing
lost in time
never to be found

our own mind
drawn to those who left
and never came back

the allure
the agony
the rumors
we can not understand
but hope to solve
one day
for them
for us

questions linger
if found
or not
what really happened
to the ones
that went missing...

MAYBE ONE DAY

Vanished
washed away
but hope
of return
maybe one day...

www.ingramcontent.com/pod-product-compliance
Lightning Source LLC
Chambersburg PA
CBHW070239220526
45465CB00004B/1456